How to Use AI to Recruit, Train and Scale A High-Performing Sales Team

Smarter Recruiting. Faster Onboarding. Bigger Results.

James Saar / Tisha Saar

Published by Honolua Publishing

Printed in the United States of America

First Edition

Table of Contents

Foreword

"AI may not replace you, but someone using AI will." — a modern truth echoed by forward-thinking leaders and echoed in boardrooms, sales teams, and recruiting circles around the world.

This quote captures the very heart of why I wrote this guide.

If you're building a team, growing a business, or simply trying to scale your impact—you can't afford to ignore the power of artificial intelligence. And the good news is: you don't have to become a tech expert to use it.

If you ever feel stuck, don't know what to write, or aren't sure where to start, just ask ChatGPT:

- "Can you give me 3 examples of prompts I could use to write a job post?"
- "What's the best way to follow up with someone who hasn't responded to my message?"

- "Help me write a welcome email for a new sales agent."

The answers you'll get won't just save you time — they'll accelerate your growth.

AI is not here to do your job for you. It's here to make you better at it.

Use this book as a playbook, a toolbox, and a springboard. Let it guide you through writing messages, building funnels, training your team, and scaling your systems—all using tools that are available to you right now.

You're not behind. You're right on time.

Let's get to work.

— James Saar

CEO, Advantage Partners

Introduction

If you've ever wished there were more hours in the day — or more hands to help you recruit — AI is about to become your new best friend.

With over **40 years in business development and sales leadership**, I've seen trends come and go—but **AI is here to stay**. This book distills the **lessons, tools, and systems** that helped me **scale teams across the country** and build **recurring income from the ground up**.

Artificial Intelligence isn't just a buzzword. It's one of the most powerful and practical tools in your recruiting toolbox. From writing job descriptions and crafting personalized outreach messages to automating follow-ups and onboarding new agents, AI can help you **build a sales team faster than ever — without sacrificing quality or human connection**.

This guide was written for **agency builders, sales leaders, and team recruiters** who want to **work smarter, not harder**. Whether you're starting from scratch or already leading a growing team, the strategies and templates in this book will help you:

- Save time by letting AI handle repetitive communication
- Reach more qualified candidates through better messaging

Build a duplicatable system for long-term growth

We're going to walk through every step of the recruiting process — from writing your first job post to building a **scalable, self-sustaining system** that practically recruits for you.

This isn't theory. These are the exact prompts, tools, and workflows I've used in the field — to grow teams, build systems, and create leverage through recurring revenue.

Let's dive in.

Chapter 1

Getting Started with AI Tools for Recruiting

If you're reading this, you're ready to build a team — and you're smart enough to want to do it faster, easier, and with more efficiency than ever before. Welcome to the future of recruiting.

Artificial intelligence (AI) has revolutionized the way businesses operate, and recruiting is no exception. With tools like ChatGPT, you no longer have to start from scratch with every job post, message, or follow-up. You can use AI to save time, improve consistency, and scale your recruiting process.

In this chapter, you'll learn:

- What AI tools are and how they work
- Where to access them (free and paid)

- Why they matter in building your dream team
- How to write effective prompts that get real results

What Is ChatGPT and How Does It Work?

ChatGPT is an AI-powered language model developed by OpenAI. Think of it like a very smart virtual assistant that can help you write just about anything — from job posts and email campaigns to training outlines and scripts. All you have to do is give it clear instructions (called "prompts") and it will generate content in seconds.

You can use ChatGPT for free at https://chat.openai.com, or upgrade to a pro version with added features. Other tools like Jasper, Copy.ai, Grok and Claude offer similar capabilities, but for this book, we'll primarily reference ChatGPT.

Free vs. Paid AI Tools

Here are a few AI platforms to explore:

- ChatGPT (Free and Pro) – Best all-around tool for writing, strategy, and automation.
- Jasper AI – Great for marketing content and team collaboration.
- Copy.ai – Fast, pre-built templates for business messaging.

- Claude – Helpful for summarizing and reworking content.

Start with ChatGPT—it's free, powerful, and easy to use.

Free vs. Paid AI Tools

Feature	Free Tools	Paid Tools
• Access to advanced AI Models	• Limited (Older modesl like GPT 3.5)	• Latest Models (GPT 4, better performace)
• Speed & Response Time	• Slower & limitd during peak times	• Faster, priority access
• Prompt Customization & Memory	• No memory or advanced settings options	• Can remember content and inputs
• Recruiting & Marketing Templates	• Create your own from scratch	• Pre-built job posts & emails
• Cost	• Free	• Typically $20-$99
• Ideal For	• Beginners, casual users	• Seriour recruiters, creators

How AI Helps with Recruiting

Here's what AI can do for you:

- Write custom job descriptions in seconds
- Generate email and LinkedIn messages
- Create onboarding checklists and training plans
- Write voicemail and text scripts
- Develop a full 12-week recruiting plan

The key? It starts with a good prompt.

Case Study: From Overwhelmed to Organized in 48 Hours

Sarah, a team recruiter in Texas, was drowning in manual follow-ups and struggling to keep her outreach organized. After learning to use ChatGPT and setting up a simple CRM + Calendly workflow, she automated her job posting and initial message replies.

"I went from chasing candidates to having candidates chase me," she said. "In the first week alone, I booked six recruiting calls without lifting a finger after setup."

Before we dive into writing prompts, take a look at the full recruiting funnel enhanced by AI. This visual will give you a quick overview of how the recruiting process flows — from the very first job post to onboarding your newest team member.

AI-Enhanced Recruiting Funnel

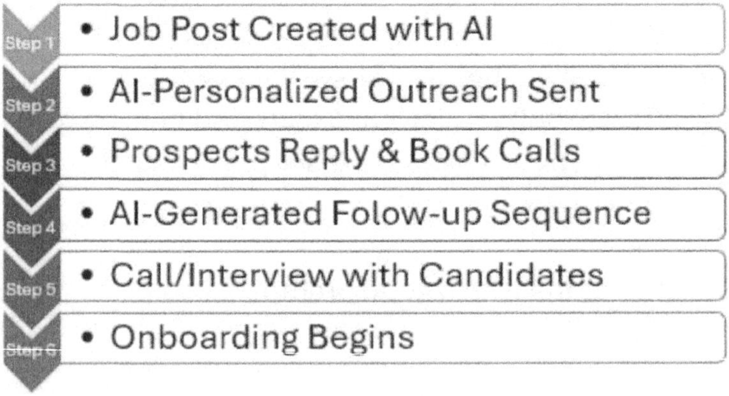

How to Write a Great Prompt

A great prompt is:

- Specific – Include the goal, target audience, and desired format
- Conversational – Ask questions like you're talking to a helpful assistant
- Adjustable – You can refine or tweak the response as needed

Example Prompt:
"Write a job description for a remote sales development rep who wants flexible hours and recurring income. Emphasize freedom, training, and team support."

ChatGPT will generate a full description you can tweak and post in minutes.

🔑 **Key Takeaways:**

- AI tools like ChatGPT can drastically reduce the time spent on repetitive recruiting tasks.
- Writing clear, specific prompts is the secret to getting high-quality content.
- Start with free tools and gradually build your AI workflow as you scale.

💬 **AI Prompts to Practice:**

- "List the top 5 ways AI can help me recruit sales agents faster."
- "Explain how ChatGPT can be used to automate recruiting without losing the human touch."
- "Compare the best free and paid AI tools for business development."

✅ **Action Steps for This Week:**

1. Sign up for ChatGPT and test at least one prompt.
2. Write a prompt and generate your first job description draft.
3. Create a simple AI command list or swipe file for future use.

Now that you've got your foundation set, let's move on to the first step: writing compelling job descriptions that actually attract the right people. That's up next in Chapter 2.

Chapter 2

Write Compelling Job Descriptions in Seconds

Recruiting starts with the job post — and a great job post does more than describe the position. It attracts the right people, communicates your values, and makes them want to learn more.

This is where AI shines.

With ChatGPT, you can write high-quality job descriptions in a matter of seconds. Whether you're looking to bring on experienced sales professionals or entrepreneurial newcomers, AI can help you craft messaging that connects with your ideal candidate.

Key Elements of a High-Converting Job Post

Here are the must-have components for a compelling job description:

- Clear, attention-grabbing headline
- Summary of the opportunity (freedom, recurring income, remote work, training)
- List of responsibilities (simple, clear, appealing)
- Benefits (flexibility, mentorship, proven systems)
- Call to action (how to apply or book a call)

AI Prompt Examples

Try these AI prompts to get started:

Prompt 1:
"Write a job post for a remote sales agent looking for flexible hours and recurring revenue. Include mentorship, training, and performance-based compensation."

Prompt 2:
"Create a job description that appeals to entrepreneurial real estate agents looking for a second stream of income with a flexible schedule."

You can even add more detail like this:

Bonus Prompt:
"Write a compelling job post for a commission-only B2B sales role that allows for remote work, offers recurring revenue, and includes weekly training and CRM access. Target experienced sales reps who are burned out from cold calling."

Template: Standard Job Posting for B2B Agents

🚀 Position: Remote B2B Sales Agent

Are you a motivated sales professional looking for flexible hours, recurring income, and a system that supports your growth?

We're looking for driven individuals who want to build long-term income by helping businesses save money using a proven, no-cost payroll tax savings strategy.

What We Offer:

- Remote, flexible schedule
- Performance-based compensation with recurring revenue
- Access to a full CRM, marketing system, and weekly mentorship
- On-demand and live training included

You'll Succeed If You:

- Have experience in sales, insurance, real estate, or business development
- Want to stop chasing one-time commissions
- Love helping businesses and building relationships

Apply now or book a call to learn more. Let's build your freedom together.

Testimonial: "I Posted and Got 17 Applications in a Day!"

"I used the job post template in this chapter, made a few small changes with ChatGPT, and posted it to Indeed and LinkedIn. Within 24 hours, I had 17 applications — and 4 of them were solid candidates. This stuff really works!"
— *Jason R., Insurance Team Leader, Georgia*

Bonus: How to Tailor Job Posts for Indeed, LinkedIn, and Handshake

- Indeed: Focus on responsibilities and salary structure. Use keywords like "remote," "commission-based," and "flexible schedule." Keep formatting clean.
- LinkedIn: Add personality and talk about mission and impact. Include a strong opening hook and end with a clear CTA to connect or apply.

- Handshake: Speak directly to students and recent grads. Highlight mentorship, training, and future growth opportunities. Use informal, energetic tone.

🔑 **Key Takeaways:**

- Your job post is the first impression — make it count with strong headlines, benefits, and clarity.
- AI can generate highly specific, targeted job posts in minutes.
- Different platforms (Indeed, LinkedIn, Handshake) require slightly different messaging.

💬 **AI Prompts to Practice:**

- "Write a job post for a remote B2B sales agent with recurring income and mentorship."
- "Tailor this job post for posting on LinkedIn vs. Indeed."
- "Add urgency and a strong call-to-action to this job post."

✅ **Action Steps for This Week:**

1. Use AI to create 2–3 variations of your job post.
2. Post on at least one platform (Indeed, LinkedIn, or Handshake).
3. Track initial engagement (views, clicks, applications).

Now that you have the job posts ready to go, the next step is getting people to read them — and respond. In Chapter 3, we'll show you how to craft personalized LinkedIn messages that open doors and build relationships with top talent.

Chapter 3

Craft Personalized LinkedIn Connection Messages

Posting a job is one thing — getting the right people to engage with it is another. That's where personalized outreach comes in. LinkedIn remains one of the most powerful platforms for connecting with experienced agents, CPAs, sales professionals, and entrepreneurial talent. But in a sea of spam and generic messages, personalization is what sets you apart.

And AI can help you do it at scale.

Why Personalization Matters in Outreach

People don't respond to copy-and-paste messages. They respond to messages that show you did your homework

— that you understand who they are, what they want, and how this opportunity can help them.

A personalized message:

- Shows genuine interest
- Builds rapport quickly
- Increases your connection and reply rates

Even with AI, you can personalize your message by feeding the tool basic details about the recipient's role, experience, or industry.

AI Prompt Examples

Prompt 1:
"Write a LinkedIn message for a CPA introducing a partnership opportunity that helps their clients save money using Section 125 and R&D tax credits."

Prompt 2:
"Draft a short message to a licensed insurance agent about joining a team that offers recurring income and flexible hours. Include a CTA to book a call."

Prompt 3:
"Create a friendly connection request for a business consultant on LinkedIn. Mention that I help business owners save on payroll taxes and increase employee benefits."

These AI prompts help you generate a variety of messages you can tweak for each outreach sequence.

Templates for Cold, Warm, and Follow-Up Messages

Cold Message Template:

Hi [First Name], I came across your profile and was impressed by your work in [industry]. I'm building a team helping business owners save thousands annually using an IRS-approved strategy—and we're earning recurring income in the process. Would you be open to a quick chat?

Warm Message Template (mutual group or background):

Hi [First Name], I noticed we're both part of the [LinkedIn Group Name] and share an interest in helping businesses grow. I'm working on a project that helps owners reduce payroll taxes and boost employee retention—would love to share more if you're open.

Follow-Up Template:

Just following up on my previous message, [First Name]. I think you'd be a great fit for what we're building. Let me know if you're open to a quick 15-minute call to explore it. Here's my calendar: [Insert Link]

Case Study: From 2% to 22% Response Rate Using AI Personalization

Vanessa, a broker in Arizona, was frustrated with cold LinkedIn outreach. By adding just two lines of personalized intro based on the person's job role and recent activity — generated by ChatGPT — her connection rate jumped from 2% to 22% in one week.

"I just gave AI a little info and let it work its magic. My messages finally sound human — and people are actually replying!"

Bonus: Using LinkedIn Sales Navigator with AI

LinkedIn Sales Navigator allows you to:

- Filter prospects by industry, role, years of experience, and more
- Save leads and build outreach lists
- Track engagement and follow-up

Pair it with AI to:

- Create messaging sequences for different prospect types
- Draft InMail messages that stand out
- Generate ideas for ongoing engagement and rapport-building

Example Prompt for AI + Sales Navigator:

"Write a three-message outreach sequence to a Sales Director in B2B who is looking for more flexibility and passive income. Keep it professional but engaging."

🔑 **Key Takeaways:**

- Personalization is critical for standing out on LinkedIn.
- AI helps scale outreach while still customizing based on role, industry, or shared interests.
- Connection, engagement, and messaging should feel authentic.

💬 **AI Prompts to Practice:**

- "Write a friendly connection message for a financial advisor interested in part-time income."
- "Generate a 3-part message sequence for LinkedIn outreach to real estate agents."
- "Create a follow-up message for someone who viewed my profile but didn't respond."

✅ **Action Steps for This Week:**

1. Send 25 personalized LinkedIn connection requests using AI-generated messages.
2. Create a basic LinkedIn message template folder (cold, warm, follow-up).

3. Respond personally to all replies within 24 hours.

Chapter 4

Follow-Up Scripts that Get Results

So you've posted your job and connected with prospects — what now?

The fortune is in the follow-up.

Timely, authentic follow-up is the difference between a missed connection and a new team member. People get busy, distracted, or hesitant. Your follow-up reminds them that there's a real opportunity — and a real person — waiting to support their next step.

And with AI, follow-up can be fast, personal, and consistent.

The Importance of Timely and Authentic Follow-Up

Waiting too long to follow up — or sending a generic message — can cost you great candidates. People appreciate personalized, prompt communication. A thoughtful follow-up shows:

- You value their interest
- You're reliable and professional
- You're willing to lead

AI allows you to scale this process without losing that personal touch.

AI Prompt Examples

Prompt 1:
"Write a follow-up message after someone expresses interest in a remote sales position but hasn't scheduled a call yet. Keep it friendly and encouraging."

Prompt 2:
"Create a reminder email for an upcoming recruiting interview. Mention the date and time, include a calendar link, and express enthusiasm."

Prompt 3:
"Write a text message checking in with someone who filled out an application but hasn't responded in a few days."

You can also give AI specific details like the prospect's name, position, or platform where they connected (e.g., LinkedIn, Indeed, Handshake).

Templates for Email, Text, and LinkedIn Follow-Up

Email Follow-Up Template:

Subject: Just Checking In – Still Interested?

Hi [First Name],

Thanks again for your interest in the remote sales position! I'd love to schedule a quick call and show you how we help business owners save money—and how you can earn recurring revenue as part of our team.

Here's my calendar: [Insert Link]
Let me know what works best for you!

To your success,
[Your Name]

Text Message Template:

Hi [First Name], just following up! I saw you were interested in our remote sales opportunity. Would love to chat and answer any questions. You can grab a time here: [Insert Link] — [Your Name]

LinkedIn Message Follow-Up:

Hey [First Name], just circling back. Are you open to a quick 15-minute call this week to learn more about the opportunity we discussed? No pressure — just want to see if it's a fit!

Testimonial: "I Didn't Realize Follow-Up Could Be This Easy"

"I've always known follow-up was important, but I never had the time or energy to stay consistent. The follow-up templates and AI prompts in this chapter helped me create a sequence that runs like clockwork. I booked three calls last week just from follow-ups I forgot I sent!"
— *Mark P., B2B Sales Consultant, California*

Bonus: Crafting Ringless Voicemail Drops with AI

Ringless voicemail is a great way to connect with prospects without interrupting their day. You can write short, authentic scripts with AI and upload them using platforms like Slydial, DropCowboy, or Slybroadcast.

Prompt:
"Write a 30-second ringless voicemail script inviting someone to learn more about a remote sales opportunity with flexible hours and recurring income. Keep it friendly and low-pressure."

Sample Script:

Hi [First Name], this is [Your Name]. I just wanted to personally invite you to check out a unique opportunity that offers remote work, training, and monthly recurring income. If you're looking for something flexible and rewarding, I'd love to chat. You can book a quick call here: [Link]. Hope to talk soon!

🔑 **Key Takeaways:**

- Follow-up is where most opportunities are won or lost.
- AI can generate personalized scripts for email, text, voicemail, and LinkedIn.
- The tone should be friendly, helpful, and timely — not pushy.

💬 **AI Prompts to Practice:**

- "Write a friendly follow-up email for someone who hasn't scheduled a call yet."
- "Create a 3-touch follow-up sequence for a recruiting candidate."
- "Draft a 30-second ringless voicemail inviting someone to learn more."

✅ **Action Steps for This Week:**

1. Set up automated follow-up emails for new prospects.
2. Record a voicemail script using an AI-generated draft.
3. Follow up with any cold leads from last week using a warm message.

Now that you know how to follow up effectively, the next step is onboarding and training your new team members. In Chapter 5, we'll show you how to use AI to create seamless onboarding flows and week-by-week training plans.

Chapter 5

Onboarding and Training New Agents

Once you've attracted and engaged the right people, your next job is to set them up for success. The faster and smoother you can onboard new agents, the more confident and productive they'll become — and the more quickly your team starts to grow.

With the help of AI, you can automate and streamline the onboarding and training process while still making it personal, structured, and easy to follow.

Using AI to Create Onboarding Sequences and Training Checklists

AI can help you build a complete onboarding experience in just minutes. From welcome emails to day-one checklists to multi-week training agendas, you can use ChatGPT or other tools to:

- Create structured training timelines
- Write emails and scripts for each onboarding step
- Develop self-paced learning plans

AI Prompt Examples

Prompt 1: "Create a 30-day onboarding plan for a new insurance agent joining a remote sales team. Include weekly goals and checkpoints."

Prompt 2: "Write a weekly training email for new recruits learning to offer the Section 125 plan. Include tips, motivation, and a call to action."

Prompt 3: "Generate a checklist for what a new agent needs to do in their first week on a sales team."

Templates for Welcome Emails, Day 1 Checklists, and Weekly Learning Plans

Welcome Email Template:

Subject: Welcome to the Team – Let's Get Started!

Hi [First Name],

Welcome aboard! We're excited to have you on the team. Over the next few weeks, you'll be learning how to help businesses save money and how to build recurring income using our proven systems.

Here's your onboarding checklist to get started: [Insert Link or Attachment]

If you have any questions, reach out anytime. You're not alone — we're here to support you every step of the way.

Let's build something amazing!

To your success,
[Your Name]

Day 1 Checklist Template:

- Set up your CRM login
- Watch the DBR Intro Video
- Review the Section 125 Training Deck
- Join the next live training call (schedule attached)
- Add 10 warm prospects to your contact list

Weekly Learning Plan Template (Week 1 Example):

Week 1 – Orientation & First Prospecting Steps

- Watch: “Section 125 Basics” video training
- Read: “How We Help Business Owners” PDF
- Task: Write your own short elevator pitch
- Goal: Schedule 2 discovery calls with business owners or brokers

To make onboarding easy and duplicatable, here’s a 30-day visual roadmap you can follow or customize for your new agents. This flow ensures every new recruit has structure, support, and momentum from day one.

30- Day Agent Onboarding Flow

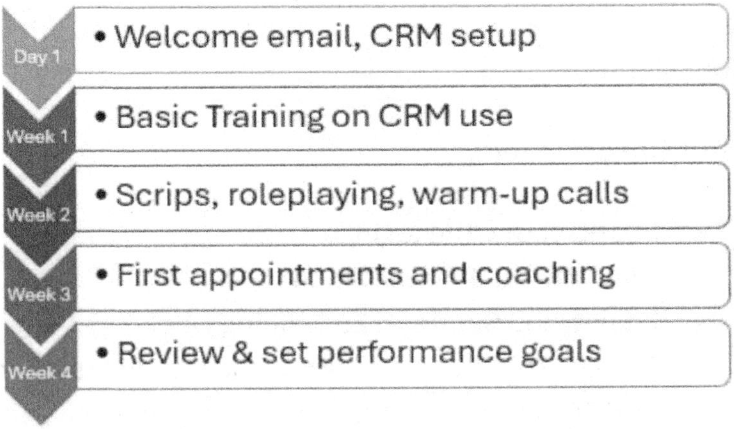

Case Study: 3 New Agents Fully Trained Without a Single Zoom Call

When Julie onboarded three agents in April, she used the AI-generated weekly learning plans, video links, and templates to guide each one through their first month.

"I didn't have to repeat myself once. The system ran itself, and all three agents booked appointments in their first two weeks. It was the smoothest onboarding I've ever experienced."

Repeat and build on this weekly with new lessons, tools, and tasks.

🔑 **Key Takeaways:**

- Structured onboarding improves retention and confidence.
- AI can help create training checklists, weekly goals, and automated email sequences.
- A duplicatable onboarding process builds scalable momentum.

💬 **AI Prompts to Practice:**

- "Create a 30-day onboarding plan for a new sales rep."
- "Write a welcome email for a new remote team member."

- "Generate a checklist for Week 1 of agent training."

✅ **Action Steps for This Week:**

1. Build or update your onboarding checklist using AI.
2. Draft 3 onboarding emails for Days 1, 3, and 7.
3. Upload materials to your CRM or training platform.

You can use AI to customize learning plans based on the agent's background (new vs. experienced) or role (recruiting vs. prospecting).

Chapter 6

Scaling with Systems and AI Integration

Building a team is powerful. But building a system that runs with or without you? That's next-level.

Once your recruiting and onboarding process is working, it's time to optimize, scale, and automate. AI and digital tools can help you create a repeatable system that grows your team faster — without burning you out.

How to Build a Duplicatable Recruiting System

A duplicatable recruiting system has three core parts:

1. A reliable flow of new leads

2. A consistent follow-up and interview process
3. A streamlined onboarding path

AI helps you build all three. The goal isn't just to recruit agents — but to recruit agents who can recruit others and follow the same system.

That means every part of your process — job posting, messaging, scheduling, onboarding — should be documentable and repeatable.

Traditional Recruiting vs. AI-Enabled Recruiting

Task	Traditional Role	AI-Enabled Recruiting
•Job Post Creation	•Manual	•Instant & custom
•Outreach Messaging	•Copy-paste templates	•Personalized messages at scale
•Follow-up	•Inconsistent	•Automated/timely
•Onboarding Materials	•Created manually per agent	•AI-generated lists, plans & emails
•Response Speed	•Slower	•Immediate
•Scalability	•Time-intensive	•Easily scaled
•Recruiting Consistency	•Varies by recruiter or team	•Standardized & repeatable system

Combining AI with Your CRM and Recruiting Funnels

Your CRM (Customer Relationship Management tool) is the engine that drives your recruiting. When you pair it with AI, you can:

- Automate messaging sequences
- Track candidate engagement
- Create AI-generated scripts for calls, texts, and emails
- Use AI to write content for your recruiting funnel (landing pages, lead magnets, email series)

Prompt Example:
"Design a 12-week recruiting plan with weekly goals and actions for someone starting from scratch."

Prompt Example:
"List 10 ways to automate lead generation for team building using free and low-cost tools."

CRM Features with and Without AI Integration

CRM Features	Without AI	With AI Integration
•Contact Management	•Manual entry, limited automation	•Auto-fill, smart data entry, tags
•Lead Scoring	•Static filters	•Predictive analytics
•Email Campaigns	•Manually written	•AI-generated content
•Task Reminders	•Manual setup	•Smart reminders
•Conversation Tracking	•Basic notes & history	•Summarized insights
•Follow-up Sequences	•Created manually	•Auto-generated interaction
•Reporting & Insights	•Pre-set reports	•AI-powered trends

Tools to Use Alongside AI

Here are essential tools to support and scale your system:

- Calendly – Automate appointment scheduling and eliminate back-and-forth texts.
- Zapier – Connect your CRM, email platform, and AI workflows to automate repetitive tasks.
- CRM Platforms – Tools like GoHighLevel, Close, or HubSpot can be integrated with email, text, and social channels.
- Notion, Trello, or ClickUp – Use these to manage onboarding tasks and track agent progress.
- Google Forms + Sheets – Capture leads, store info, and track KPIs without expensive software.

Testimonial: "Finally, I Have a Recruiting Machine That Runs Without Me"

"Between Zapier, my CRM, and ChatGPT, I now have a real system. I get notifications when leads book themselves, onboarding content is auto-sent, and all I have to do is show up to support. This chapter was a game-changer."
— *Darren W., Agency Owner, Colorado*

Final Tip: Document Everything Once

Every email you write... every message that gets a response... every checklist or follow-up sequence you build—save it, name it, and systematize it.

Then share it with your team.

Scaling is about systems, not superheroes.

🔑 **Key Takeaways:**

- Systems—not hustle—create long-term scale.
- AI plus CRM tools can automate almost every stage of recruiting.
- Documenting your process ensures anyone can follow it and succeed.

💬 **AI Prompts to Practice:**

- "Design a 12-week recruiting plan with goals and action steps."
- "Write a content plan for a recruiting funnel with email, social media, and video scripts."
- "List tools I can use to automate appointment booking, follow-up, and lead nurturing."

✅ **Action Steps for This Week:**

1. Document your entire recruiting flow (job post to onboarding).
2. Set up one automation (e.g., Calendly + email reminders).
3. Use AI to draft a basic 3-email recruiting funnel.

In Chapter 7, we'll wrap things up with a 90-day Freedom Launch Plan — so you can take action and start seeing results immediately. It's time to scale with automation and smart structure.

Chapter 7

Your 90-Day Freedom Launch Plan

You've made it this far — so you already know more than 99% of people who want to build a business, but never take action. The truth is, everything you need to start growing your team, creating leverage, and generating recurring income is right here in front of you.

The final piece? Taking bold, focused action — consistently.

That's what this chapter is all about: turning strategy into momentum and helping you launch your recruiting machine with confidence.

Why the First 90 Days Matter Most

The first 90 days set the tone for everything that follows. It's the sweet spot where energy meets execution — and where smart habits can compound into major results.

You don't need to be perfect. You just need to be in motion.

The next 90 days will lay the foundation for long-term success. Use the visual roadmap below to follow the key phases of building your recruiting machine. Treat this like your GPS — don't skip steps, and celebrate each milestone along the way.

90-Day Recruiting Launch Roadmap

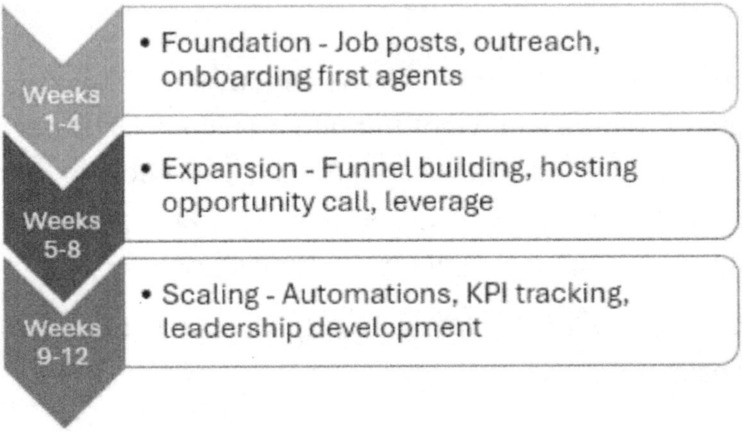

Your 12-Week Launch Plan

Weeks 1–4: Foundation and First Wins

- Set up your CRM, Calendly, and LinkedIn profile for success
- Post your first job description using ChatGPT-generated templates

- Reach out to 50–100 prospects using LinkedIn and Indeed
- Schedule 3–5 discovery calls per week
- Onboard your first 1–2 agents using your new checklist

Weeks 5–8: Expansion and Leverage

- Refine your messaging with AI based on responses
- Record a short recruiting video for follow-up and ads
- Launch your first automated email sequence or funnel
- Host a live or recorded opportunity call
- Invite your first team members to co-recruit with you

Case Study: Launched, Leveraged, and Doubled His Team in 90 Days

Michael, a sales trainer in Florida, committed to the 12-week plan. He scheduled weekly prospecting blocks, let AI handle job posts and messages, and recorded a simple "why join us" video.

"In 90 days, I went from 2 active agents to 5 producers, and now I spend more time coaching than recruiting. This plan gave me my time back — and my income went up."

Weeks 9–12: Systems and Scale

- Document your repeatable process (onboarding, outreach, training)
- Use ChatGPT to create social media posts, drip emails, and scripts
- Plug new agents into your training content and weekly calls
- Set your monthly team recruiting goal and track KPIs
- Celebrate your first recurring commission deposits—and teach your agents how to do the same

Tools and Tips to Keep You on Track

- Use Calendly to book calls without back-and-forth
- Use ChatGPT daily to brainstorm ideas, draft messages, and save time
- Track your activity weekly: messages sent, calls booked, recruits enrolled
- Reuse your best content—record once, repurpose everywhere
- Stay connected with your upline and your team—this is a relationship business

🔑 **Key Takeaways:**

- Consistent action over 90 days beats bursts of effort.
- This chapter gives you a structured, momentum-building roadmap.
- Progress, not perfection, is the goal—track activity and adjust.

💭 **AI Prompts to Practice:**

- "Break down the 12-week plan into weekly checklists."
- "Create a motivational weekly message for my team."
- "Write a milestone celebration email for agents completing their first 30 days."

✅ **Action Steps for This Week:**

1. Commit to your 90-day recruiting launch.
2. Print or copy the 12-week plan and check off each week.
3. Share your launch goal with a mentor or accountability partner.

Final Words of Encouragement

You don't need to know everything. You just need to start.

The systems are here. The tools are ready. The opportunity is real.

Whether you're doing this part-time or full-time, whether you've built a team before or this is your first time — you can do this.

The best time to build recurring revenue was yesterday. The next best time? Right now.

Go build your dream. Go build your legacy. Go build your freedom.

Conclusion

Recruit Smarter, Lead Stronger

Throughout this book, you've learned how to harness the power of AI to recruit smarter, grow faster, and build a team with purpose. From writing job descriptions and crafting personalized outreach to automating onboarding and scaling your systems — you now have everything you need to launch and lead with confidence.

AI isn't here to replace you — it's here to empower you. It gives you time, clarity, and leverage. It helps you show up consistently, communicate effectively, and support your team without burning out.

But even the smartest tools mean nothing without action. The first step is always the hardest — and also the most important.

So take it.

Appendix A

AI for Team Retention & Motivation

Most AI recruiting books stop at onboarding. But if your goal is recurring income and long-term growth, then *retention* isn't optional — it's essential.

The truth is, building a team is only half the battle. Keeping that team **motivated, connected, and producing consistently** is how you create real leverage.

That's where AI becomes your secret weapon — not just for recruiting, but for team culture and communication. You can use ChatGPT to create weekly motivation messages, recognition scripts, and even ideas for contests, training sessions, and newsletters — without spending hours every week reinventing the wheel.

Why Retention Matters in a Recurring Revenue Model

- High turnover = high time cost (more recruiting, more training)
- Happy, engaged agents = better performance, more referrals, and stronger team culture
- Culture compounds — and automation can help you scale it

💬 **Weekly Motivation Message Templates**

Use ChatGPT to generate fresh motivation each week. These messages can be posted in team chats, emailed out, or included in your onboarding and training flows.

✉️ **Template Prompt:**

"Write a weekly motivation message for my sales team. Make it encouraging, mention teamwork, and include a reminder to book at least 2 calls this week."

📥 **Example Output:**

Subject: 🔥 New Week, New Wins
Hey Team,

Just a quick reminder: this week is a fresh start, and every conversation brings you one step closer to your goals. Stay focused, keep it fun, and remember — you're not doing this alone.

Set your intention today: Book 2 calls, follow up with 3 leads, and take one step outside your comfort zone. Let's make it count!

— [Your Name]

Recognition Scripts (for Calls, Emails, or Team Posts)

Public recognition is one of the most powerful motivators — and you don't have to write these from scratch every time.

🎤 **Prompt:**

"Write a short recognition message for a new agent who just booked their first discovery call. Keep it energetic and encouraging."

📝 **Example:**

Huge shoutout to **Alicia** for booking her first discovery call this week! 🎉 Taking that first step takes courage — and it shows you're ready to grow. Keep going, Alicia — we're cheering you on!

More Recognition Prompts:

- "Create a thank-you message for a team member who referred a great candidate."
- "Write a celebratory message for an agent who just earned their first recurring commission."

AI-Generated Ideas for Team Contests & Culture

Use AI to brainstorm creative and engaging ideas that build momentum and community.

🎯 **Prompt:**

"Give me 5 fun and motivating sales team contests to boost recruiting activity for remote agents."

💡 **Example Output:**

1. **The 5-Call Challenge** – Everyone who books 5 calls this week gets entered into a drawing for a $25 gift card.
2. **Motivation Monday Quotes** – Submit your favorite motivational quote and win a prize if yours gets the most votes.
3. **Tag-Team Tuesday** – Partner up and challenge each other to hit weekly KPIs.
4. **LinkedIn Legend** – Post about your opportunity on LinkedIn. Best post wins recognition + swag.

5. **Rookie of the Month** – Celebrate the top-performing new agent with a featured shoutout and reward.

Quick Team Newsletter Generator

Stay top of mind and keep your team informed with a short, engaging newsletter — written in seconds.

Prompt:

"Write a short weekly team newsletter that includes a win of the week, upcoming training, and a motivational quote."

Example Output:

Team Weekly Recap – Week of July 10
Win of the Week: Congrats to Jon S. for signing 2 new clients this week — your consistency is paying off!
Upcoming Training: Join us Thursday at 3 PM PST for a special training on overcoming objections (Zoom link below).
Quote of the Week: "Success is the sum of small efforts repeated day in and day out." – Robert Collier

Final Tip: Automate and Reuse

Once you've created a few great messages or contest templates, don't start from scratch next time. Use AI to tweak and refresh content, then **batch schedule** emails or team messages weekly using your CRM or email platform.

Consistency builds culture. And AI makes consistency easy.

Summary

Retention is leverage. Motivation is fuel. AI is your system.
You don't need to be a cheerleader every day — you just need a plan, a few prompts, and a weekly rhythm. Start using AI not just to grow your team — but to keep them *inspired*, *engaged*, and *thriving*.

Appendix B

Recruiting Progress Tracker & Scorecard

Success in recruiting isn't about perfection — it's about **consistent action** and **measurable progress**.

This Recruiting Progress Tracker & Scorecard is your weekly accountability partner. Whether you're building your team part-time or running a full-scale recruiting machine, this simple tool helps you stay focused on the activities that actually drive results:

- Messages Sent
- Calls Booked
- New Agents Onboarded
- Weekly Revenue Earned

When you track your activity, you build **momentum**. You spot trends. You see what's working. And you stay motivated — because every message, every call, and every new agent brings you one step closer to the freedom and income you're building.

Use this scorecard weekly to:

- Set realistic goals
- Measure your consistency
- Celebrate your wins
- Identify areas for improvement

Weekly Scorecard

Week #	Your Goal	Actual
• Messages Sent	• ________	• ________
• Calls Booked	• ________	• ________
• New Agents Onboarded	• ________	• ________
• Weekly Review	• ________	• ________

Weekly Activity Summary

Day	Messages Sent	Calls Booked
• Monday	• ________	• ________
• Tuesday	• ________	• ________
• Wednesday	• ________	• ________
• Thursday	• ________	• ________
• Friday	• ________	• ________

Weekly Reflection Goals

Biggest Win of the Week:

- ____
- ____
- ____

What I'll Improve Next Week:

- ____
- ____
- ____

New Scripts, Strategy or Tool I Tried:

- ____
- ____
- ____

Goals for Next Week

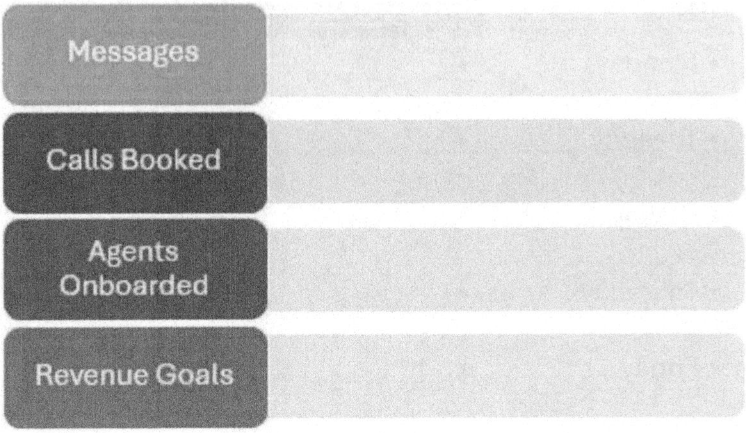

Whether you print it, save it digitally, or use it as part of your CRM routine, make this tracker a part of your rhythm. Because what gets **measured** gets **multiplied**.

Let's build your dream — one conversation, one call, one agent at a time.

Appendix C
AI Prompt Library by Recruiting Stage

Use these ready-made prompts with ChatGPT or your preferred AI assistant. Customize as needed — your goal is to give clear intent and context for each output.

🔍 1. Recruiting & Job Posts

1. "Write a job post for a remote B2B sales role with recurring income, training, and flexible hours."
2. "Summarize the benefits of joining a team that offers Section 125 plans to business owners."
3. "Write a job description that appeals to entrepreneurial real estate agents looking for part-time income."

4. "Create a job posting for new or part-time insurance agents seeking flexible, remote opportunities."
5. "Write a concise job ad for a Handshake listing targeting recent college grads with an interest in finance and business."
6. "Write a job description for a commission-only role that emphasizes mentorship, training, and long-term income potential."
7. "Create a job listing for a virtual inside sales role that highlights recurring commissions and team support."
8. "Write a compelling Indeed ad for a remote sales agent helping businesses save on payroll taxes."
9. "Create a high-energy job post to attract entrepreneurial moms or dads looking for remote work."
10. "Write a Craigslist-style ad in a friendly, casual tone for a part-time sales agent who loves helping small businesses."

💬 **2. Outreach & Messaging**

These prompts generate cold and warm connection messages, sequences, and personalized communication at scale.

1. "Create a LinkedIn connection request message for an insurance agent looking for more flexibility and recurring income."
2. "Write a cold DM for a financial advisor about joining a referral partnership."
3. "Generate a 3-part outreach sequence to engage business consultants with a recruiting offer."
4. "Write a connection message for a real estate agent looking to add another income stream."
5. "Create a short, informal Facebook message to reconnect with a former colleague about a business opportunity."
6. "Draft an InMail message for a sales leader who recently posted about burnout or job dissatisfaction."
7. "Write a warm intro message for someone in a mutual LinkedIn group focused on entrepreneurship."

8. "Create a connection message tailored to CPAs or bookkeepers, offering them a new way to help their clients."
9. "Write a follow-up message for someone who liked your post but didn't respond to your invite."
10. "Generate a short email to invite someone to a virtual opportunity call about building residual income."

📞 3. Follow-Up

Use these prompts to create professional, friendly, and effective follow-up messages across channels.

1. "Write a follow-up email for someone who expressed interest but didn't book a call."
2. "Create a text message reminder for a recruiting call happening tomorrow at 1 PM."
3. "Draft a friendly voicemail for a missed connection following a job inquiry."
4. "Write a message to check in with someone who filled out a form but hasn't responded in 3 days."
5. "Create a 'just checking in' email for someone who received onboarding info but hasn't started."

6. "Draft a second follow-up for someone who said they were interested but went quiet."
7. "Write a short message thanking a candidate for applying and asking if they have any questions."
8. "Create a message to re-engage a former agent who went inactive last year."
9. "Write a message reminding a candidate about a scheduled Zoom interview and what to expect."
10. "Draft a follow-up text asking if someone is still interested in learning more about the team opportunity."

📚 4. Onboarding & Training

These prompts help automate and personalize onboarding, training, and team development communications.

1. "Create a 30-day onboarding plan for a new remote sales agent."
2. "Write a welcome email for a new agent joining my team, including login info and day 1 checklist."
3. "Draft a motivational weekly email for new recruits focused on booking discovery calls."

4. "Generate a Week 2 email with prospecting tips, link to recorded training, and encouragement."
5. "Create a checklist for a new agent's first 7 days including CRM access, training videos, and outreach goals."
6. "Write a short daily reminder email for the first 5 days of onboarding to help new agents stay on track."
7. "Draft an email announcing a live training session and why it's important to attend."
8. "Create a template for a weekly 'Agent Spotlight' to recognize new team members."
9. "Write a sample onboarding message that can be sent via text after someone is approved to join."
10. "Generate a script for a welcome video message introducing yourself and the next steps."

Appendix D
AI Glossary for Non-Tech Savvy Readers

A quick-reference guide to help non-technical users feel confident with AI language and tools.

AI (Artificial Intelligence)	•Technology that simulates human thinking, like problem-solving or writing.
ChatGPT	•An AI tool developed by OpenAI that generates human0like text based on prompts.
Prompt	•The instruction or request you give to the AI to produce a result.
Language Model	•A trained system that understands and creates text. ChatGPT is on example

GPT-3.5 / GPT-4	•Versions of the AI model; GPT-4 is more advanced, used in paid plans
Free vs. Pro	•Free AI Tools often use older models and have limits; Pro versions are faster and more capable
Tokens	•Units of text used by the AI to calculate response length. One word = 1 token
Fine-Tuning	•Training an AI model on specfic data or tone. Most users don't need this - prompts are enough
Integration	•Connecting AI tools with your CRM or calendar so tasks happen automatically.
Automation	•Setting up AI to perform tasks (like follow-ups or reminders) without manual input.

Appendix E
Recommended Tech Stack for AI-Enhanced Recruiting

Building a scalable recruiting system doesn't require a computer science degree — just the right tools, used the right way.

This appendix outlines a **simple, effective tech stack** that supports every stage of your recruiting journey: from posting jobs and automating outreach, to onboarding new agents and tracking team performance. Whether you're a tech novice or a seasoned builder, these tools will help you **save time, stay organized, and scale faster** — without sacrificing the human connection that makes your team strong.

You don't need to use every platform on this list. Start with the essentials, add as you grow, and focus on **function over flash**. The goal isn't to have more software — it's to build **smart systems** that work for you.

Each tool below includes a quick description, pros and cons, and a direct link so you can explore what fits your needs best.

Let's simplify your tech so you can focus on building your dream team — with clarity, confidence, and automation.

AI Tech Stack

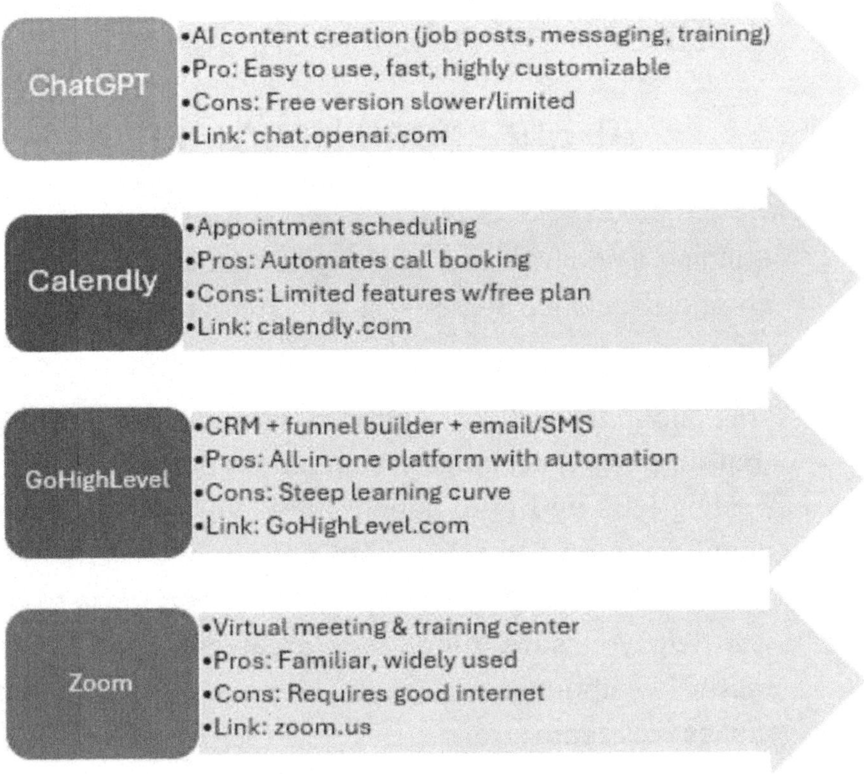

Trello / Notion /ClickUp

- Task & contect management
- Pros: Free version , visual, team friendly
- Cons: May require team onboarding
- Link: notion.so / trello.com

Zapier

- Automation between apps
- Pros: Connects your CRM, email, calendar, and more
- Cons: Can get expensive to scale
- Link: zapier.com

Slybroadcast

- Ringless voicemail drops
- Pros: Saves time, personal touch
- Cons: Legal limitations in some states
- Link: slybroadcast.com

Google Forms & Sheets

- Lead capture + tracking
- Pros: Free, easy to use, customizable
- Cons: Not as robust as CRM
- Link: forms.Google.com

Appendix F

Your Opportunity to Build Lifelong, Recurring Income with Your Team

---

If you've made it this far, it's clear—you're serious about creating a business that doesn't just pay once, but pays you **over and over again**.

DREAMS Business Resources (DBR) is more than a system — it's a complete business-building platform designed to help you **recruit, train, and scale a sales team**, while generating **residual, long-term income month after month after month**.

The DREAMS model was built for **entrepreneurial leaders, agency builders, and sales professionals** who want to create leverage and financial freedom — not just make one-time commissions.

At its core, DBR is a **plug-and-play ecosystem** where everything you need to grow is already built for you — from tools to training to mentorship.

What DREAMS Has for YOU:

- **D**uplicate Systems
- **R**ecurring Revenue
- **E**ducation & Empowerment
- **A**utomation
- **M**entorship
- **S**cale

What You Get with DREAMS Business Resources:

Done-for-You Systems & Tools

- Pre-written **job posts**, **AI messaging templates**, **scripts**, and **funnels**
- Access to a fully integrated **CRM** (track leads, automate messages, onboard new agents)
- Built-in follow-up sequences to keep prospects and agents engaged

Weekly Live Coaching & Support

- Weekly **group coaching calls** led by **international sales trainer Eric Lofholm**
- Real-time support, roleplay, and Q&A

- Learn what's working *now* from other top producers and mentors

🎓 **On-Demand & Step-by-Step Training**

- Easy-to-follow onboarding for new agents
- Recruiting and sales training available 24/7
- Training paths for both **builders** and **producers**
- **Mentors** for 1-on-1 coaching

🤝 **Collaborative Community**

- Connect with like-minded leaders, coaches, and business builders
- Access private groups, referral partners, and accountability circles
- Build a business **with others**, not by yourself

💰 **Why DREAMS Works:**

Most people struggle to grow a team because they're trying to do everything manually — job posts, follow-ups, interviews, onboarding — all from scratch.

DREAMS Business Resources gives you a **proven, duplicatable system** that anyone can follow. It simplifies recruiting, **amplifies your reach with AI**, and automates the heavy lifting — so you can focus on relationships and results.

The result? A growing team. A scalable system. And recurring income that comes in whether you're actively selling or not.

🔑 **Who It's For:**

- Licensed insurance agents and financial pros
- Entrepreneurs looking for passive income opportunities
- Leaders who want to **recruit and scale a national team**
- Real estate agents, network marketers, or professionals transitioning into remote work

Whether you're brand new or already building, this system was designed to help you **scale faster**, **recruit smarter**, and **build real freedom**.

While DREAMS is a powerful platform for team building and recruiting, it's also a **high-value B2B sales solution** for professionals who want to help business owners **save money, improve employee benefits, and increase profitability** — all with **zero cost to the business**.

If you're looking for a way to add massive value in the B2B space, DREAMS gives you the tools, training, and strategies to **become a trusted consultant**, not just a salesperson.

✅ Business Solutions Offered Through DREAMS:

DREAMS isn't just a recruiting system — it's a powerful, done-for-you B2B sales platform that gives you everything you need to bring real, measurable value to business owners across the country. These solutions are designed to improve profitability, reduce expenses, enhance employee benefits, and strengthen operations — all while helping you earn residual monthly income.

Section 125 Payroll Tax Savings Plan

- Businesses save $500–$900 per employee, per year in payroll taxes
- Employees receive a net pay increase, enhanced wellness benefits, and zero out-of-pocket cost
- No disruption to existing health plans — and no cost to implement
- Fully insured, IRS-approved, and available in all 50 states

R&D Tax Credits

- Helps businesses unlock thousands to hundreds of thousands of dollars in federal R&D credits
- Especially impactful for manufacturers, farms, wineries, tech companies, construction firms, and engineers

- Zero risk: Fees only apply if the business successfully secures credits
- Backdated for up to three years, plus potential for quarterly returns

🏥 Supplemental Health Indemnity Plans

- A zero-cost, optional employee benefit program
- Includes hospital indemnity, accident, critical illness, mental wellness, telemedicine, and even pet care
- Boosts morale and employee retention with no cost to the employer
- Includes coaching, Rx plans, and wellness perks that are often more valuable than traditional coverage

🏦 Business Bank Loans with No Credit Requirements

- Offer businesses access to capital without requiring:
    - A minimum credit score
    - A personal guarantee
    - Collateral
- Funding available in as little as 24–48 hours
- Ideal for growth, debt consolidation, equipment upgrades, or cash flow stabilization

Energy Cost Reduction & Commercial Solar Programs

- Help businesses reduce utility expenses through solar partnerships and energy audits
- Offer access to zero-upfront-cost solar options for qualified commercial properties
- Includes proposals for power purchase agreements (PPA) and long-term energy savings
- Great for environmentally conscious businesses looking to increase their sustainability

And So Much More...

DREAMS is constantly expanding its platform to include new value-based solutions for business owners. As a DREAMS Partner, you'll be equipped with:

- Exclusive access to vetted vendors and programs
- Pre-built presentations and scripts to support B2B conversations
- Ongoing training on how to approach and consult businesses across multiple industries
- A complete system for growing your B2B pipeline with tools that deliver results — not just pitches

💼 How DREAMS Supports B2B Sales Professionals

DREAMS provides a full **sales enablement platform** for those who want to grow a book of business with monthly recurring revenue. Whether you're selling one solution or all four, you'll get:

- **B2B Prospecting Scripts** tailored to each vertical (e.g. CPA firms, clinics, construction companies)
- **Email and LinkedIn outreach templates** generated and optimized using AI
- **Done-for-you sales tools**, including presentations, client decks, proposal templates, and benefit summaries
- **CRM to track business leads**, follow-up, and client conversions
- **Training on objection handling, gatekeeper strategies, and consultative selling**

💰 Why Business Owners Say Yes:

- There's **no upfront cost**
- The process is **simple, fast, and compliant**
- They **save money**, improve benefits, and **retain staff**
- They gain a **trusted partner** who brings real value — not just another pitch

🤝 Built for Independent Reps, Consultants, and Agencies

DREAMS is ideal for:

- Insurance agents & brokers
- Financial advisors
- HR consultants
- Tax pros & CPAs
- Anyone looking to build **B2B relationships** and earn **recurring income**

📌 **Bottom Line:**

DREAMS isn't just about recruiting a team — it's about serving businesses with high-value solutions that pay you month after month.

You get everything you need to grow a B2B pipeline:

- Proven offers
- Automation tools
- Training and support
- A brandable, compliant platform

And best of all? You're helping business owners **win** — while building a business of your own that runs on **real value and real results.**

Next Steps

📍 Need more information? Visit: **DREAMS Business Resources at:** https://umustsee.net/CPX6GR

✅ Ready to join the team? Go to: **DREAMS Partner Enrollment at:** https://dreamsresources.com/join/?refid=AA3778

📅 Book a 15-minute strategy call and let's launch your recurring income plan together. Here is a link to my calendar: https://calendly.com/jsaar/ai15

This is your moment. Your freedom formula. Your launchpad.

Recruit smarter. Lead stronger. Build your dream. It's time.

Made in the USA
Monee, IL
04 November 2026

67377428R00056